P9-BHR-843

A WISH YOU WERE HERE© BOOK

WINDOWS OF THE PAST

RUINS OF THE COLORADO PLATEAU

FLORENCE C. LISTER & LYNN WILSON

ISBN 0-939365-21-9 (PAPER)
0-939365-22-7 (CLOTH)

First printing 1993
Printed in Singapore by Tien Wah Press, Ltd.

Title Page Photo:
CORNER WINDOW, CHACO CULTURE N.H.P., N.M.

SIERRA PRESS, INC.

PO BOX 430, EL PORTAL, CA 95318

CONTENTS

ACKNOWLEDGEMENTS

We wish to thank the following individuals, many of whom are employees of the National Park Service or their Natural History Associations, for their assistance, inspiration and support during the long and arduous process involved in bringing this book to press: Kristie Arrington, Robert M. Baker, Jan Balsom, Tina Begay, John E. Cook, Nancy Coulam, Walt Dabney, Dale Davidson, Larry Davis, David Dunatchik, Rovilla Ellis, Don Fiero, Ron Foreman, Dave Forgang, Pam Frazier, Larry Frederick, Larry Fulfer, Elaine Giles, James Hakala, Joy Hilding, Bruce Hucko, Alice Johns, Jose Knighton, Victor Knox, David Krouskop, Nicky Leach, Janet Lowe, Robert Mack, Bruce Mellberg, Jim Miller, Tami Morton, Theresa Nichols, Herschel Schulz, Kim Sikoryak, Shirley Torgerson, Mark Wellman, Eva Wells and especially Florence!

DEDICATION

This book is dedicated to our parents,
With love.
And to parents everywhere,
Teach your children well.

INTRODUCTION

This book is about ruins.

These are the ruins of the architecture of two different cultures known today by their Spanish and Navajo names: Sinagua and Anasazi, respectively. I cannot tell you what these people called themselves.

It is about Cliff Palace and White House and Montezuma's Castle and it is about Keet Seel and Chettro Ketl and Wukoki. It is about the facts and the illusions.

This book is about fabulous cities of stone and it is about an inconspicuous granary tucked into a sandstone alcove fifteen miles up an unnamed canyon somewhere west of nowhere in southeast Utah.

It is about the Sinaguan sites of Lomaki, Wupatki and Tuzigoot, but I cannot tell you what the Sinaguans called them,

and it is about the Anasazi sites of Hovenweep, Betatakin and Hungo Pavi, but I cannot tell you what the Anasazi called them.

This book is about what we know (you must read Florence Lister's magnificent essay—"The Prehistoric Drama"); and it is about our shared humanity (you must read Lynn Wilson's intuitive five-part poem—"Windows"); and it is about how these ruins appear to a number of very talented photographers (all photo credits appear on page 96).

Lastly, this book is about respect; for the people responsible for building these pieces of a priceless Native American heritage; for the ruins and rock art sites themselves, be they sacred or secular; and finally, respect for today's descendents of these early Americans. Treat these fragile sites with respect and reverence and they will be here for our grandchildren and for the great, great, great...grandchildren of the original inhabitants.

THE PREHISTORIC DRAMA

by Florence C. Lister

ENTER THE PALEO-INDIAN

First came the hunters. Out of the mists of Siberia, tiny bands ventured eastward across the land bridge of the Bering Strait and ever deeper into a territory stretching from pole to pole, where no human being had trod before. It was a momentous migration that would affect the history of mankind, but it is doubtful that the participants had any sense of the extraordinary. As they penetrated farther inland, their casual coming was prolonged into permanence at the end of the Ice Age with the rising of sea waters, which resubmerged the land connection between the Eastern and Western hemispheres. So here they were.

Generation after generation, successive waves of men and women rolled southward, pulled by the tidal force of large mammals whose land this was. By some fourteen thousand years ago, possibly even earlier, the migrants had settled around the bog-filled basins and grassy prairies of the southern section of North America. Human history dawned over the Southwest and the Colorado Plateau, which would much later be the scene of an unfolding saga, the leading characters of which were a people we know today by the Navajo name of Anasazi.

The early hunters were few in number, and the landscape into which they dispersed was vast and constantly evolving, often violently. Their baggage of material possessions, the stuff of archaeology, was meager: traces of their passage limited to a few scattered spearpoints, stone and bone implements used in food or hide processing, flecks of charcoal or fire-reddened rocks from ancient hearths, and the bruised bony residue of their kills and butchering. They may have made wooden, skin, or basketry articles as well as temporary brush huts. Remains of the individuals themselves are virtually absent from the record. While it is generally presumed these Paleo-Indians were modern men of Mongoloid racial

THE GREAT GALLERY, CANYONLANDS N.P., UT

stock, as yet there is no proof. They remain a physical enigma.

One thing is certain, however, these Paleo-Indians were people of intelligence. They demonstrated consummate skills as flint knappers and specialized hunters, arts they had refined after long practice. They propelled tools of their trade—large lanceolate, fluted points of fine-grained stone attached to long wooden spears—with the aid of *atlatls*, or spear throwers. Dexterity was essential in handling this awkward device in order to secure sufficient thrust to penetrate vital parts of the huge thick-skinned or furry beasts upon which they preyed. It took the coordinated efforts of teams of hunters to down one animal, which typically outweighed a normal man by many hundreds of pounds. On occasion they drove hapless creatures into traps, spongy marshes, over precipices, or plucked off sick strays.

Kill sites must have presented a scene of feverish activity. It was here that the people dismembered the carcasses of their victims and smoked, dried, or pounded them with animal fat into pemican. In winter they may have frozen the meat, perhaps even caching it until spring. They cracked bones for marrow and stretched and scraped the hides to be used as coverings.

Over the following three or four millennia the elephant-like mammoths and the camels, sloths, and horses upon which these first American residents subsisted slowly died off. Whether these mammals were on a downward path toward extinction when the Paleo-Indians arrived or whether they were exterminated by human predators is not known. But what happened to the hunters? Did they vanish with their customary food supply? Did they adapt to new circumstances? Or did another influx of people moving out of Canada bring a slightly different technology of small tools and new life blood?

We know that a second category of animal life filled the vacuum left by the disappearance of the first. Perhaps Folsom people succeeded Clovis people in a similar fashion. Whatever the precise sequence of events, those groups now known as Folsom were as skilled as their Clovis predecessors in stone projectile point manufacture and in hunting. Apparently they also were more numerous, especially along the eastern flanks of the greater Southwest, where for thousands of years lush grasslands had attracted thunderous herds of bison. But when cyclical climatic change dried up the vegetation and water holes, the Folsom complex faded. It either dissipated entirely or was absorbed into other similar hunting-based societies, who themselves were either unsuccessful or moving into a new cultural stage regional scholars term the Archaic.

A TIME OF TRANSITION

The Archaic period began a prolonged segment of time when many of the dominant themes of life in this region were introduced. The people who were responsible for this unappreciated, but critical, transitional scene were dependent upon both food procurement ways of the past and newly emerging techniques: they were hunters and gatherers.

From all directions, Archaic folks drifted into the Southwest in the form of loosely organized, extended families or small social units who moved seasonally across this landscape in a share-and-share-alike quest for food. They camped in the open or under the protection of overhanging ledges, cooking over shallow rock-ringed depressions, bedding down in patches of cleared ground. In addition to an assortment of small stone spearpoints used to kill game animals and ingeniously devised snares and nets to trap rabbits, prairie dogs, and rats, the refuse they left behind reveals an especially significant use of vegetal materials. Desiccated plant parts and pollen reflect the dietary importance of such foodstuffs, which were ground upon stone grinding slabs using handstones—the ubiquitous *metates* and *manos* found throughout the Southwest. Although, from time to time, the Ice Age hunters before them

TWIN TOWERS, HOVENWEEP N.M., UT.

likely supplemented their meaty meals with seeds, greens, nuts, and tubers, such items became a fundamental food group for Archaic peoples. Rice grass, amaranth, purslane, tansy mustard, beeweed, prickly pear cactus, sage, piñon nuts, and squawberry enlivened the diet as primary ingredients and as seasonings.

The variety of stone used for implements; of bones left from the hunt; of fibers processed for new handicrafts, such as the weaving of cordage, sandals, and baskets; and of plants used as medicines or dyes suggest an impressively expanding data base about the physical world in which the Archaic groups daily operated. For those in the northern half of the greater Southwest, that world was the Colorado Plateau.

The Colorado Plateau is an uplifted area thousands of square miles in size that encompasses parts of modern Arizona, New Mexico, Colorado, and Utah (now known as the Four Corners). It is a region of extreme topographical, ecological, and climatic diversity. The range within each of these categories is so dramatic as to make much of the land at opposing ends of their spectrums seem unsuitable for human occupation. Yet, there are few localities on the Colorado Plateau without evidence of former human presence, from the rare instances where evidence of early big game hunters may be found to clear traces of the later Archaic nomads.

Despite the environmental challenges, the Archaic bands found a rich inventory of flora and fauna to exploit. Doubtless, their dependence upon Nature made them sensitive to the rhythms of the seasons, of sky, of earth, and of greater forces in a universal scheme. Supplications may have been made to the latter for success by pecking on to cliff faces petroglyphs of mountain sheep, deer, or antelope; they placed small split-twig figurines of similar animals in special spots as ceremonial offerings ; and archeologists have found at least one rock art panel representing a harvest scene of figures using sickles and seed-beating tools to gather ripened grasses. Larger-than-life pictographs (painted rock art) of armless anthropomorphs with trapezoidal bodies and rudimentary heads hover ghost-like in remote rocky galleries—evidence of an expanding world view and that these people employed shamans to conjure spirits to ensure their well-being.

Archaic bands were in the Southwest when the first of several waves of Mesoamerican influence spread northward. This wave introduced the revolutionary idea of agriculture, primarily the raising of maize. Some thousands of years earlier maize, or corn, had been hybridized from a native grass to become such an important staple for formative Mexican societies that it was regarded as a sacred gift of the gods. At the second millennium B.C. some Archaic men, perhaps already knowledgeable through experimentation in matters of wild plant propagation, growth, and maturation, saw the advantage in helping the processes along. They planted kernels of corn, received by way of long-distance trade channels, into holes poked in hillocks of loosened earth and returned months later to reap the rewards, if any. It may be that periodic drought conditions that withered native vegetation underscored for them the security that might be possible with purposeful nurturing, or it may be that population increases had outstripped undomesticated reserves. Whatever the reason, the idea gradually took hold. Hunter-gatherers were becoming hunters, gatherers, and gardeners. Three major cultural entities (the Hohokam, the Mogollon, and the Anasazi) and several lesser groups emerged from this background, each adapting to divergent southwestern environments. Upon the Colorado Plateau, the Anasazi left the most significant mark.

A CULTURE IS FORMED

Centuries passed before the idea of farming led to appreciable changes in

WALL DETAIL, SALMON RUINS, N.M.

the life-style of people living on the Colorado Plateau. Maybe some of the people thought that cultivating a garden required excessive labor in an area only marginally suitable. Others among them may have found their nomadic life-style ran too deeply in them to be abandoned. Nevertheless, the prospect of stocks of homegrown produce eventually won them over. The Anasazi cultural stage was ushered in by at least 1500 B.C., and the subsistence strategy based upon agriculture was to mold it.

PEOPLE OF THE SOIL

The heartland of the Colorado Plateau, where the Anasazi culture germinated, bloomed, and faded, is split by physiographic features into three provinces. While the native peoples in each shared a basic orientation, they also made unique adaptations to their differing environments.

To the north of the San Juan River, which flows westward out of the lofty tangle of the southern Rockies to merge with the Colorado River at Lake Powell, is what archaeologists refer to as the Mesa Verde district in reference to the forested mesa closely identified with the earliest archaeological finds in the area. It is the most verdant of the three zones and is generally higher in elevation, blessed with deep soils, and, in its eastern sectors, a mountain watershed. As the western portion of the district breaks down to the San Juan and Colorado rivers, it is rent by abrupt craggy canyons and panoramas of colorful bare rock pinnacles and buttes.

South of the San Juan River and east of the Chuska-Lukachukai mountain ranges that form the north-south divide between New Mexico and Arizona is the Chaco district. This is the San Juan Basin, a huge waterless depression stranded between the normal winter storm path out of the northwest and the summer storm path from the southeast. Chaco Canyon, actually a shallow valley, cuts generally east-west across the basin. A scant tree cover is found only on the crests of a few mesas. Over a bleak encirclement of low sage, hundred-mile vistas tug a viewer in all directions.

The Kayenta district, which encompasses today's Navajo and Hopi Indian reservations, stretches south of the San Juan River from the Chuskas west to the Colorado River. Its most famous natural features are Canyon de Chelly in the east and Monument Valley in the west. This region is dry below about 7,000 feet, with shallow soils that repeatedly boil up into clouds of wind-borne red dirt. There are few perennial sources of water.

To farm with some degree of success in any of these provinces required a new knowledge of the environment. Types and locations of productive soils, the complex balance of elevation and temperature, the length of growing seasons, storm tracks, and water sources all became a part of the pooled tribal survival wisdom.

The first Anasazi farmers, known to archaeologists as the Basketmaker II according to a cultural chronology (the Pecos Classification) accepted by the profession in 1927, devised a simple set of tools to carry out the arduous task of clearing and planting virgin land. These were stone axes, stone-bladed hoes, long stout sticks, and baskets in which to haul off unwanted dirt and rocks. There can be no denying that such primitive farming technology involved hours of backbreaking toil.

Much Anasazi farming was of the dry-land variety, not unlike that practiced in the same regions by white settlers arriving in the late nineteenth and early twentieth centuries. Garden plots were situated at mouths of gullies, where they were moistened by runoff from summer showers, on terraces by watercourses to take advantage of floodwaters, or on high mesa tops, where chances were better for a steadier rate of

PETROGLYPH, PETRIFIED FOREST N.P., AZ.

year-round precipitation. As experience increased, farmers tried a variety of means to manage water and soil. They placed lines of rocks across channels or on slopes to form check dams, either to cause enriched water-laid deposits to build up or to capture drainage. They dug irrigation canals to bring intermittent or perennial water to fields. They erected small earth and rock dams and reservoirs. More projects on a scale large enough to require cooperative efforts were undertaken in the Mesa Verde and Chaco districts than in the Kayenta, where population density was less and the possibilities for farming fewer.

At first, corn and squash were the only two vegetables raised. During the sixth century A.D., beans (*Phaseolus vulgaris*) were added to the roster. They were probably planted close to the corn so that its stalks could serve as trellises. Some time later, in the warmest climates, cotton began to be grown. Although not a food, its fibers were important in the weaving of textiles. Additionally, some desirable untamed plants were allowed to colonize gardens.

Meanwhile, the search for wild meats and plants continued. The more efficient bow and arrow was introduced and replaced the atlatl. Turkeys, whose feathers were woven into blankets and clothing, were domesticated.

HANDPRINTS, PICTOGRAPHS, SOUTHEAST UTAH.

A thousand years into this great agricultural revolution, a sophisticated dual environmental response evolved in arid Chaco Canyon. It illustrates how skilled the farmers of the high desert had become. Gardeners working along the south side of the canyon put their plots on the numerous talus fans at the base of the bounding cliffs where, because of their northern exposure, they absorbed late spring snow melt and where, because of sandy ground, they retained moisture from summer rains. Theirs was a typical Anasazi practice.

However, less than a mile away, across the canyon, where there are no talus slopes at the foot of the cliffs, a different tactic was required. That cliff escarpment is vertical and is topped with exposed rock extending back at least a quarter to a half mile, which slopes southward toward the canyon. Periodic violent summer storms drop hundreds of thousands of gallons of water on this rock rim which have no place to go but over the edge. Anasazi farmers living below the north escarpment of Chaco Canyon learned to harvest these waterfalls by diverting them into canals and reservoirs, and then directing the water by means of ditches and headgates to gardens bordered with earthen berms. It was a form of floodwater farming without benefit of a river system; more than likely the intermittent Chaco Wash in the canyon bottom already was too entrenched to be usable in this fashion. The rush of water over exposed sandstone brought down minerals that enriched the bottomlands, as well as a quantity of sand, which served as mulch. In time, this constant flooding would lead to salinization, which would threaten these undrained, gridded gardens. Prior to that setback, however, two very different farming techniques were functioning simultaneously on opposite sides of the canyon. During good years in the A.D. 1000s the nine-mile length of Chaco Canyon — given over in modern times to sagebrush, greasewood, and rattlesnakes — must have been a green oasis.

ANASAZI HOME BUILDERS

Successful tending of crops meant staying near them, at least during growing season. It also meant finding a way to store surpluses of dried corn, squash, and beans to be used as seed and food reserves. Thus, it was farming which dictated the development of the first Anasazi architecture.

Most of the fledgling farmers chose to settle in the rugged canyon country of the Four Corners. Here, numerous alcoves honeycomb the multihued stony ramparts of the Mesa Verde and Kayenta archaeological districts and offer protection from the elements. Other farmers among them preferred the foothills of the San Juan Mountains or the more open settings of the northern Rio Grande valley. Lacking deep canyons with rock shelters, the Chaco district held no appeal for settlers of this time.

Initial attempts at construction focused upon hoarding foodstuffs rather than creating personal shelter. Shallow storage cists were hollowed into the soft dirt fill of alcove floors. These pits often were either lined with stones arranged upright around the perimeters or were coated with sealants of thick mud. Stone slabs sometimes set on supporting logs served as cist lids to keep the contents clean and out of reach of hungry rodents. Cists not employed to store edibles occasionally functioned as convenient places to store the dead, who, in the prevailing dry atmosphere, became mummified. It was in such places that desiccated bodies were first encountered by excavators fifteen hundred years later. They named them the Basketmakers because of numerous artifacts of this type left as grave goods. Unlike the Paleo-Indians, these individuals later could be linked to the racial strains of modern Pueblo Indians.

The cists of the Basketmakers may have inspired them to experiment with similar constructions to house themselves. Early pithouses were essentially the same sort of rudimentary circular depressions but of larger circumference, with upper walls and a roofing of logs, brush, and earth. Usually a central opening in the roof above a hearth in the floor below served as both a smoke hole and an entrance. Less often, access was through a side opening. Dark, stagnant, and constricted, such pithouses probably were intended only for sleeping, with most daytime activities taking place outside. Created from the natural materials at hand and nestled into the ground, the pithouse can be seen as implying a growing Anasazi feeling of oneness with the earth around them.

Once the harvest was over, the Basketmakers likely loaded their few worldly possessions on their backs and took off on a seasonal round of gleaning Nature's bounty. From time to time some of their number returned to the caches left in alcove floors, but whether in spring the groups as a whole resumed residence in the same places and reworked the same patches of ground is uncertain. It may be that a pattern of mobility already established in the Archaic heritage also typified the Anasazi lifeway, with extended pauses attributable to the demands of farming. When thin soils were depleted of their nutrients or when the next valley looked more promising, the people moved with no apparent trauma. Maybe it was relocation just a few miles away, but much of Anasazi building was temporary and occupied for perhaps no more than several decades. In an endless cycle, shelters were built and then vacated to tumble back to the earth from which they arose. It was as though there was no enduring attachment to a particular spot, but one's sense of place came from the rootedness in a plateau vastness fading to the horizon.

Optimum weather conditions, a more abundant and nutritious diet, and the arrival on the plateau of peoples from surrounding areas meant a growing Anasazi population by A.D. 500. Numerous hamlets of ten or so pithouses and surface work areas sprang up on mesa tops, river terraces, and in valleys where there had been no earlier occupation. By this time, all three provinces of the Colorado Plateau were part of the

PICTOGRAPH, ARIZONA STRIP, AZ.

Anasazi realm. The pithouse became a more standardized domicile. It was larger and dug more deeply than before, with embellishments such as mud-plastered walls, an interior encircling bench cut from the soil, a vertical side shaft from floor to surface level to provide fresh air, a central fire pit shielded from ventilator drafts by an upright stone slab, an ash pit, occasional low earthen partitions to demark special function zones, wall pegs on which to hang objects, and floor cists. A truncated roof of logs and earth was supported on four posts set into the earthen floor in a squared pattern and lashed with split yucca ties to connecting cross beams. Sometimes uprights were added to rectify trouble overhead. Insulated by the surrounding ground, the pithouses were warm cocoons in winter and cool caves in summer. Most villages also included a pit structure of greater dimensions than the family dwellings that served as a community center for secular and sacred affairs.

Several centuries later Anasazi builders took their chances above ground. They erected linear rows or crescents of attached, flat-roofed, rectangular rooms with walls of closely spaced vertical poles sealed in mud, a style of building known today by its Spanish name of *jacal.* A secondary rear row of comparable but smaller units was built to stow stocks of domesticated and wild foods for winter consumption. Together, these small apartments comprised the first of the communal houseblocks that became characteristic of the remainder of the Anasazi and, beyond that, of the contemporary Pueblo era. Whether living side by side separated by shared walls was a matter of conservation of resources and energy, or not, communistic attitudes, in this instance reflected in their architecture, are what drove the Anasazi social system.

Typically, in front of the jacal dwellings, there was a retention from the past: a pit structure, which archaeologists now call a *kiva.* Generally it was circular, but some western Anasazi in the Kayenta district had a preference for square kivas. Based on analogy with modern pueblos, the kiva is thought to have been a sacred sanctuary embraced in the womb of Mother Earth. It always contained a symbolic floor hole opening to the spirit world, the *sipapu,* from which the Anasazi believed themselves to have emerged in legendary times and through which one day they would return.

While some Anasazi continued to dwell in pithouses and jacal row houses, especially those in the Kayentan western parts of the plateau, others upgraded their basic architecture. By the late 900s, as the Anasazi culture was approaching its apex most Anasazi lived in compact, multiroomed masonry buildings housing perhaps as many as a half dozen nuclear families. Such edifices were thickly spread over the Mesa Verdean and Chacoan districts and accommodated a burgeoning populace, although the Kayentan district experienced less growth. Cobblestones or ashlars of sandstone laid in copious amounts of mud mortar were the usual materials for contiguous cellular rooms, which were mud-plastered inside and out. Flat roofs were created by layers of primary and secondary rafters, secured across wall tops in opposite directions and covered with shredded bark, stone slabs, and earth. The only apertures were low, narrow doorways, probably closed with mats, hides, or shaped sandstone blocks. Floors were of packed earth. Kivas often were sunk into the ground in front of the house, or, if part of the structure, were made to appear subterranean.

Regional differences in the usual building formula were due primarily to available supplies and quality of workmanship. Gradually, local

PETROGLYPH, NORTHERN ARIZONA.

idiosyncrasies evolved: towers, kivas within towers, particular kiva enrichments, super-sanctuaries called Great Kivas, jacal partitions, earthworks, or circular units formed by two or three concentric walls and cross-partitions. The use of local resources and the blocky geometric elevations that mirrored those of the landscape created an aesthetically pleasing visual harmony between shelter and setting.

Masons at Chaco Canyon were responsible for the most outstanding architecture of the period. The enormous amount of building that took place in Chaco during the tenth and eleventh centuries reflects the same distinctions as were evident in the differing agricultural styles practiced by north-side and south-side residents. The south-side traditionalists built hundreds of village houses one story in height, consisting of ten to twenty small, nondescript rooms strung out in single or double rows. They used unshaped sandstone blocks crudely cushioned in a great deal of mud for walls, which were then covered with mud plaster. Two or three kivas placed within the room complex served the populace. Living space must have been simple but adequate.

Meanwhile, north-side innovators exhibited tour de force skill, surging far beyond anything attempted by their contemporaries. Without the aid of mechanical means, they created great houses of hundreds of rooms, which rose from the canyon floor, some stacked in tiers along the back wall up to four stories high and reached by external ladders. Dozens of residential kivas were incorporated whose flat roofs provided work space. Every detail was preplanned and formalized, although executed in increments. The floor plans of most of these huge edifices were horseshoe-shaped and open toward the south to trap heat reflected off the cliffs behind them. Enclosed plazas soaked up winter sunshine. Frequently within these courtyards were one or more specialized Great Kivas, where the entire community could have come together for particular rituals. These kivas were mammoth in comparison with other constructions, ranging from forty to sixty feet in diameter. They were always sunk into the ground at least ten to twelve feet but had raised walls to allow entrance through surface antechambers. They contained floor vaults that may have been used as foot drums, a firebox, an encircling bench, wall niches, and four mighty columns to hold up a roofing superstructure weighing many tons.

Although the same basic construction supplies of stone, earth, and wood were used in the Chaco great houses as were used in other contemporary works, they were special in type and kind or in the immense effort needed to obtain and use them. Dark sandstone from strata on the north mesa of Chaco was quarried with stone mauls and axes and fractured along bedding planes into billions of neat tablets of varying thickness. These were laid in several coursed styles and served as veneer for a rubble wall hearting. Tiny rock spalls stuck between masonry layers appear decorative but were there as levelers and helped strengthen the bond between mortar and stone. Gallons of precious water must have been consumed in mixing sufficient mud mortar to hold this masonry in place. Was it caught during summer rains, or were settling ponds dug in the sands of Chaco Wash? Hefty pine and spruce timbers, sixteen to twenty feet in length, spanned rooms of extraordinarily large dimensions for Anasazi buildings and permitted ceilings eleven feet in height. A veritable forest was felled to raise the roofs of the dozen such structures within Chaco Canyon and its immediate vicinity. Stands of such timber were from forty to ninety miles distant.

This compulsion for monumental construction by one segment of

DOORWAYS, CHACO CULTURE N.H.P., N.M.

Chaco society did not stop there. A far-flung scattering of some seventy-five houseblocks, of smaller size and made from immediately available stone and wood, now has been identified throughout the San Juan Basin and northward into the Mesa Verde district. A statement of authority as expressed through architecture is implied.

Building structures more than one story in height was a daring Chacoan experiment probably tried in part because of the needs and rewards of agriculture. Six to eight hundred rooms up to four or perhaps five stories high, as at the impressive site of Pueblo Bonito, obviously encroached less upon limited tracts of arable land than if the rooms had been all on one level. Additionally, many of the interior ground floor cells exhibit no evidence of having been used as dwellings. It is plausible that, with access through a single door or a hatchway in the ceiling, they were well-protected vaults for raw materials, craft goods, and foodstuffs. If so, their size and number may partially confirm the gardening success of either Chacoan farmers or others contributing to the stockpiles. The stones of some interior rooms remain burned from intense ancient fires, but whether the conflagrations were due to accidents, arson, pillage, or spontaneous combustion remains an intriguing question.

To buttress the argument for upper stories being used as dwelling space and lower stories for storage, some students of Chacoan culture point to similar living arrangements at some modern Hopi towns and at Taos Pueblo. Conversely, other researchers are convinced that daily living activities took place only in ground-level rooms and, furthermore, that the occupancy rate of the great houses at any given time was far less than the number of rooms suggests. Residency, in fact, may have been seasonal. In this interpretation, more usual population estimates of about fifty-two hundred individuals is reduced to between one and two thousand. Admittedly, at this time it is an argument to which there are no decisive answers.

Another type of construction, related to the Chacoan agricultural base, and an economic, possibly spiritual, system that evolved as part of it, was hundreds of miles of roads connecting Chaco Canyon to distant satellite communities, or outliers. These roads were cleared paths uniformly thirty feet in width, with stones or berms piled alongside as curbing, and were obviously engineered to run straight across the terrain. They went through or over obstacles, rather than winding around them, and connected to broad stairways cut into the north and south sandstone escarpments on either side of Chaco Canyon. Since the Anasazi had neither wheeled vehicles nor draft animals, one suggestion about the roads' function was that they served to facilitate human transport of goods from the outlying communities united in a trade alliance. In addition to construction timbers, raw materials for handcrafted household objects or implements, and finished goods, food stuffs in short supply at particular times may also have been carried along these routes. With extreme seasonal variability in the storm track across the San Juan Basin, there was an ever-present threat of crop failure in some sectors while others enjoyed good harvests. In this theoretical model, a communistic imperative underlies the road system. Other cultural reconstructions have the roads serving as formalized routes over which pilgrims traveled to the Chacoan great houses for ceremonial events or as being merely nonfunctional symbolic ties between specific great houses and far-flung colonies.

After the Chacoan great houses were abandoned, about A.D. 1150, most were allowed to collapse into heaps of stone and dust. The final

WINDOW & WALLS, CHACO CULTURE N.H.P., N.M.

Anasazi architectural developments of note occurred during the thirteenth century, primarily in the Mesa Verde district. A very sizable population, then north of the San Juan River, gathered into a number of sprawling one-story settlements clustered around springs at the heads of canyons or were placed in parallel rows of adjoining rooms across valley swales. Two or three-story towers became a feature of towns at the edge of the Great Sage Plain of southwestern Colorado, as seen at places such as Hovenweep National Monument. It is possible that these towers were astronomical observatories where priests could calculate the movements of heavenly bodies and so arrange critical agricultural calendars. Some towers were connected to kivas by underground passageways. Perhaps, because emergence is a dominant Puebloan cosmological theme, they were part of the stage setting needed for rituals by which costumed participants could appear as apparitions from the spirit world. An inordinate ratio of kivas to rooms exists in some places, such as at Yellowjacket and Sand Canyon pueblos. A few examples contain white-plastered lower walls painted with black geometric patterns and floors engraved with Kokopelli (flute player) fertility symbols. The powerful imagery suggests intensification of religious anxieties such as might have been aroused by a natural world seemingly gone awry.

ANASAZI CLIFF DWELLERS

The most puzzling development of the times was a mass movement back to cliff recesses, where the Anasazi architectural sequence had begun a thousand years earlier. The thirteenth-century builders were undaunted by the difficulties and dangers of access and the prodigious construction efforts required to use them. This time, instead of just storage cists, the entire town complex as it had evolved over the centuries—rooms on top of rooms, kivas, dance plazas, granaries, and turkey pens—was placed within high alcoves along strata seams and on naturally cantilevered rock ledges. Structures were molded like putty to the configurations of the selected locales and seemed to emerge from the rock in which they huddled. Occasionally, retaining walls held back dirt brought in to increase usable space. Rooms were put on top of or around fallen boulders too ponderous to move. Cave roofs were room ceilings. The ruins of these dwellings (classic examples of which are Cliff Palace or Long House in Mesa Verde National Park and Betatakin and Keet Seel in Navajo National Monument) are often situated in spectacularly beautiful but improbable surroundings and littered with artifacts from the past. It was scenes such as these that first ignited the imaginations of the American public at the end of the nineteenth century and launched the study of southwestern prehistory.

A great number of cliff dwellings and storage rooms cling to the walls of canyons within the confines of Mesa Verde National Park. Two years after the chance sighting in 1888 of the largest site of Cliff Palace, one hundred and eighty others of varying size were reported in the vicinity. Many more were found in the rough country of southeastern Utah, still within the range of Mesa Verde cultural expression. In spite of this concentration of cliff structures in the northern sectors of the Colorado Plateau, it is not certain that they were solely a local response to some unknown conditions. In the 1200s, when some Mesa Verdeans were on the move into localities lacking canyon topography, such as the shallow tributary valleys of the upper San Juan drainage or Chaco Canyon, they reoccupied and remodeled vacated flatland buildings. When they arrived at Canyon de Chelly and Canyon del Muerto in northeastern Arizona, however, they took advantage of

'TEXTILE' PETROGLYPH, WUPATKI N.M., AZ.

rock shelters in the sheer cliffs and, as at Mummy Cave, built directly on top of Basketmaker cists. Mesa Verdean influence also may have inspired construction of cliff houses in the Kayenta district during the last quarter of the thirteenth century. Betatakin, Keet Seel, and Inscription House, while Mesa Verdean in appearance, are clearly Kayentan in workmanship. At present, the best that can be said is that by unknown processes and for obscure reasons, hundreds of fellow Anasazi simultaneously developed a taste for elevated secluded real estate.

The concept of perching houses and storage units on cliff frontages was not confined to those Anasazi on the Colorado Plateau but was diffused from them to their contemporaries: the Mogollon, Salado, and Sinagua peoples occupying the more southerly Southwest or the western periphery of the plateau. Gila Cliff Dwellings in southern New Mexico and Tonto, Walnut Canyon, and Montezuma Castle national monuments in Arizona are examples of this borrowed trait. Whether the southerners were faced with the same potential threats as their neighbors or were merely taking up a new fashion is not known.

HANDPRINT PICTOGRAPHS, NATURAL BRIDGES N.M., UT.

A PENCHANT FOR BEAUTY

If Anasazi men created a unique architecture that is the enduring hallmark of their culture, Anasazi women were responsible for the two equally distinctive crafts of basketry and pottery making. Examples of both have survived because they were frequent grave furnishings. Records of the total output, however, are uneven because the products of one craft are perishable and fragments of the other are almost indestructible. Even so, it is obvious that after mastering the techniques of manufacture, latent aesthetic urges surfaced that resulted in totally individualistic statements of the artisans' world.

On the Colorado Plateau, basketry preceded pottery by several thousand years. It evolved in the Archaic milieu because of the need for lightweight portable containers as the bands moved across their collecting territories. When emphasis on nomadic gathering continued into the emerging horticultural complex, craftswomen practiced five procedures—twining, twilling, plaiting, wickerwork, and coiling—to turn out functional objects of many sizes, shapes, and textures. Material life grew more elaborate. String bags and flexible burden baskets were used for cartage. At meal times, hot roasting rocks were swirled with gleanings in large basket trays up to two feet in diameter, while other trays were employed for winnowing and storage. Tightly woven jar forms dabbed with pitch were used as water bottles.

Although decoration of these containers was irrelevant to their function, it was not long before the weavers began to experiment with design. They did this by developing a new technology of dyeing splints in solutions made from pulverized minerals or plants. The resulting black, red, yellow, and blue-green splints created patterns interwoven into the natural fiber background. Weaving techniques restricted motifs to geometric patterns of bands, spirals, and radials of considerable complexity which were arranged with arithmetic precision. The balance and dynamism exhibited in these design fields, in addition to their outstanding craftsmanship, show a remarkable progression over two or three centuries from a minimal artistic sensibility to one of a high order.

Pottery making became a major cultural impulse that diffused northward from central Mexico reaching the Anasazi during the fifth century A.D. The Anasazi's location at the northern limit of Mesoamerican influence and their strong basketry tradition slowed acceptance of an idea already embraced by southern neighbors such as the Mogollon and the Hohokam. Once adopted, however, pottery became indispensable in every household for storage, cooking, and serving. It went hand in hand with the commitment to greater sedentism. Thus, farming made pottery a practical solution to domestic needs.

The first several centuries of pottery production were ones of great experimentation in technology and style. Since the women who made baskets likely also became potters and the end products were meant for the same purposes, there was an inevitable transfer of ideas from the established craft to the new one. Earthenware pots were begun within basket bases and were then built up by concentric ropes of clay in the construction manner of coiled baskets. The first pots were small round-bottomed jars that could be balanced in cookfire embers. Soon shallow bowls imitated serving baskets. Their interior surfaces, which would never be fire-blackened, invited decoration. This was accomplished by means of yucca fiber brushes and pigments comparable to those intended for dyeing basket elements. Basketry was the inspiration for early decoration. An isolated central circle in bowl bottoms recalls the initial coil of some baskets. Stitch patterns and some geometrics likewise appear derived from the basket repertory. Occasional human stick figures, while paralleling both petroglyphs and pictographs, also hint at a happy liberation from past basketry tradition.

The critical firing process, which converts earth to earthenware, was part of the package of ceramic skills borrowed from Mogollon peoples to the south straddling the diffusionary path from Mexico. Pots were stacked on the ground and covered with fuel, which was allowed to burn freely. This created an oxidizing atmosphere that baked the vessels to a reddish-brown color. Shortly thereafter, Anasazi artisans began to bank their fires so that oxygen was reduced. This left the finished pots a dirty gray. Decorations painted on before firing turned black.

By about A.D. 700, Anasazi women had settled into methods and a decorative format that would persist with few modifications for the next six hundred years and that would make their wares distinctive in the aboriginal New World. Plain-surfaced gray vessels continued to be used for cooking and storage. Greenware to be decorated was coated with an engobe, or slip, of fine-grained clay that fired white to contrast dramatically with the black designs. Modern ceramicists and archaeologists have had only limited success in consistently duplicating the white background color, the tone quality, and the adhesiveness of the decorative pigments, yet there is little evidence that the Anasazi experienced many failures. They knew about resources and methods still unknown to present scholars.

Ceramic forms tell of the increasing elaboration of Anasazi lifeways. Canteens, water jars, ladles, and pitchers held or dispensed liquids. Bowls of many sizes were used for food preparation and serving. The increase in number of these specimens or pieces left from breakage is indicative of a vastly expanding number of makers and users.

Although some plain gray wares continued to be made throughout the Anasazi continuum, an unusual surface finish called corrugation was introduced as the culture matured that dominated gray utility ware for the rest of the Anasazi residence. Exterior coils were left unsmoothed to facilitate handling and to increase the thermal properties of vessels by exposing more clay surface to the heat. The desire to alter this texture surface aesthetically is shown in the precise pinching of coils, sometimes into diagonal, curved, or horizontal patterns. Functional advantages were not increased by this time-consuming decorative attention, but the pots were handsome to look at and must have been

WINDOW, TWO-STORY RUIN, GRAND GULCH, UT.

satisfying to the craftswomen.

Painted designs focused upon a dozen or so geometric motifs used repeatedly but arranged without any explicit duplication. They evoked the angularity of the surrounding landforms, the primal elements, the intensity of light and shadow. Subtle symbolism was inherent.

During the same period that regional variations became apparent in architecture, corresponding differences occurred in pottery, as artisans unwittingly drew on their experiences as inspiration. Those potters in the Chaco district specialized in fine-line hatchure, deftly executed in a mineral pigment, and often laid in hooked bands diagonally across a chalky-white design field. The visual effect was a lightly grayed, torqued pattern of great energy. If, as some researchers believe, Chaco Canyon's great houses comprised an ostentatious ceremonial center to which scores of pilgrims trudged, it may also be that their tall, straight-sided cylinders and hollow figurines of animals or humans were intended for showy rituals. A century later, Mesa Verde decorators preferred a bold, banded layout composed of large, glossy, black painted elements balanced with white spacing. This kind of design could only have been conceived by an ordered, conservative society. Characteristically Mesa Verdean forms were small, handled mugs and squat, ovoid jars with lids. They may have been paraphernalia for intimate clan or family kiva functions. In addition to black-on-white wares, Kayentan potters made red types that often displayed polychrome decoration. The production of red pottery may be a clue to lingering connections with the Mogollon tradition. Kayentan classic black-on-white design modes emphasized the use of black to achieve a negative patterning in white. There were no forms of probable ceremonial purpose, just as there were few kivas. Kayenta potters turned out excellent pots, their expertise undoubtedly resurfacing in later Hopi ceramics.

Artistic merit is seen in a variety of lesser goods, whether destined for mundane or special use. Pendants, beads, ear bobs, and fetishes of stone, bone, or shell; turquoise inlays and mosaics on stone, shell and basketry; carved and painted wooden staffs and altar pieces; and textiles woven from plant and animal fibers all display a talent for making something beautiful and original out of the humblest of materials.

THE OUTMIGRATION

Even after a century of scrutiny, the Anasazi remain elusive in many ways. Perhaps that is fitting because the region in which they dwelt also defies complete comprehension. Crumbling silenced houses and museum vaults filled with artifacts take cultural reconstruction only so far. Without a written record, the remainder of the chronicle comes from inference, from speculation, from analogy with descendants and their material goods and world views.

Nothing is known of Anasazi languages, although it is presumed there were several. Dances, songs, oral traditions, body painting, and most particularized and everyday costuming are blanks in the accounting. Social and political ordering can only be speculated upon, based on public works and architecture. Knowledge of religious ideology is indirectly found in rock art, burials, seemingly ritualized constructions, and aesthetic expressions that show no separation between the secular and the sacred. Trading connections have been revealed in identifiable commodities, some from as far away as northern Mexico and the Gulf of California. The staying power of the Anasazi is tantalizing testimony of their success in coping with one of the continent's most high-risk environments, but there were obvious negatives.

GRANARY DOOR, GRAND GULCH, UT.

Some of these negatives are found in the physical remains of the Anasazi themselves. They reveal a high infant mortality rate, with few adults living beyond forty years. Broken bones, tooth decay, and untended maladies were common. The rigors of a close-to-earth life-style and illnesses due to the spread of bacteria from communal habitations, burials within occupied quarters, lack of daily sanitation, and poor nutrition took their toll. Stressful conditions brought on perhaps by enemies within and without may explain palisaded villages, enclosed great houses, sealed lower wall apertures, dwellings in cliff faces or on sheer-sided ridges, and occasional signs of violent deaths and cannibalized human bones. Prehistoric life in the Four Corners region of the Colorado Plateau was never idyllic; by A.D. 1300 it had become impossible to continue without dramatic change.

In the decades leading up to that time, the physical world of the Anasazi became out of tune with its environment. Some of the troubles were man-made. The land they revered was raped through ignorance. Denudation, salinization, exhaustion, erosion, and over-exploitation by a population density in some sectors not since matched brought despair, if not panic. These problems were compounded by shifts in precipitation patterns, which caused prolonged droughts. It is likely that internal dissension over the perceived failure of religious leadership, possibly even intertribal raiding, followed. Chaco Anasazi living in the least favorable agricultural zone were the first to flower and the first to wither. They took the traditional Anasazi course of action: they moved. Many probably drifted eastward to the Rio Grande valley. Others went southeast toward Acoma or southwest toward Zuni. Others may have dispersed into twelfth-century outliers, where they were absorbed by northern San Juan Anasazi. Most of those people at home in the hospitable Mesa Verde district hung on for another century, but during that same time large numbers of them also began pulling up stakes and migrating southward along the upper San Juan River, over into the Rio Grande valley, briefly into Chaco Canyon, and out into northeastern Arizona. In that latter district they met up with Kayentan companions, who themselves were struggling for survival. The die-hards left in the north congregated into large settlements, paid increased homage to Mother Earth and Father Sky, and tried to maintain the status quo. To no avail. As the fourteenth century opened, the Colorado Plateau, home for at least several thousand years to a people determined to control their future through agriculture, was left to heal itself. During the next several hundred years the centers of population continued to shift southward and eastward. In Arizona, uprooted Anasazi from both the Mesa Verde and Kayenta districts gravitated into large communities that spread from the southern base of Black Mesa to the Little Colorado drainage to the Mogollon Rim. Gradually they concentrated in the vicinity of the Hopi mesas. In New Mexico, former Chacoans and Mesa Verdeans settled in the regions about Zuni and Acoma and along the central to northern Rio Grande valley, making up the nineteen pueblos found in that state today. When the Spaniards arrived in the mid-sixteenth century, they called these native peoples the Pueblo Indians because they lived in towns, or pueblos. That name persists, but their cultural and racial ancestry is one that was evolved over centuries by the Anasazi of the Colorado Plateau.

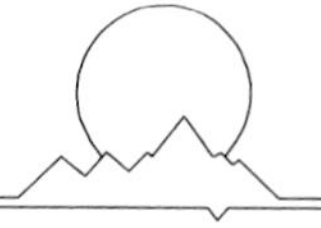

QUAIL PANEL, SOUTHEAST UTAH.

WINDOWS

Lynn Wilson

WINDOW, PAIUTE CANYON, NAVAJO RESERVATION, UT.

DAWN

Luminous Thunder Egg of night,
Moves from shadow into light.
Morning stars soon fade away,
Awakens now the eye of day.

Go to sleep nocturnal things,
Scamper out daylight beings.
Add a mystic twilight hue,
Gentle breeze, prism'd dew.

Softened light, morning prayer,
The scent of ages fills the air.
Men too gaze, with much delight,
This glowing dawn of golden light.

There was oneness with each day,
Prayer and ritual, every way.
A diligent quest to understand,
The complex workings of his land.

There was no I, from what i see,
Each task was simple harmony.
Body energy that was spent,
Had purpose for which it went.

One thing's the same, this is time,
As was then, so tis mine.
Rising sun, warmth of day,
Sounds of laughter, children play.

Much the same, the sky above,
Falling rain, a mothers love.
And so this dawn of Ancient Man,
Has much to do with who i am.

This moved me to spend one day,
Experiencing life as the Ancients may.
Let earthly rhythm embrace my being,
Smell, touch, sound, as well as seeing.

SPRUCE TREE HOUSE, MESA VERDE N.P., CO.

KIVAS AT PUEBLO BONITO, CHACO CULTURE N.H.P., N.M.

BETATAKIN, NAVAJO N.M., AZ.

KEET SEEL, NAVAJO N.M., AZ.

WHITE HOUSE, CANYON de CHELLY N.M., AZ.

CASA RINCONADA, CHACO CULTURE N.H.P., N.M.

HORSECOLLAR RUIN, NATURAL BRIDGES N.M., UT.

WUPATKI RUIN, WUPATKI N.M., AZ.

WEST PUEBLO, AZTEC RUINS N.M., NEW MEXICO.

UNNAMED SITE, GLEN CANYON N.R.A., UT/AZ.

REBUILT PIT HOUSE, STEPHOUSE SITE, MESA VERDE N.P., CO.

GRANARY, CHETTRO KETL, CHACO CULTURE N.H.P., N.M.

FAR VIEW, MESA VERDE N.P., CO.

HOVENWEEP HOUSE, HOVENWEEP N.M., UT.

THUNDERSTORM, TUZIGOOT N.M., AZ.

RUIN IN ROAD CANYON, CEDAR MESA, UT.

MID-DAY

PUEBLO BONITO, CHACO CULTURE N.H.P., N.M.

Suns aloft, dry heat of day.
Sweat drips, on earthen clay.
Fetch the water, grind that corn.
Plow those fields, babies born.

Stacking stones, sealing holes.
Painting walls, setting poles.
Around the crops children play,
Keeping brother bug away.

Drawn symmetry of this day,
Decorates a bowl of clay,
Or perhaps on cloth that's sewn,
Laced with needles made of bone.

Molding pots layer by layer.
Singing songs of ancient prayer.
Rainbows fill a blackened sky.
A scent of ozone, a storm is nigh.

Lightning candles shadowed land,
i watch with glee from where i am.
Thunder shakes dry, heated air.
Raindrops fall on earthen bare.

Rocks echo, creak and groan.
Plants are born, seeds are sown.
Desert land swallows deep.
Drink up cacti and mesquite.

Red-tailed hawks soar humid air.
Winds play with midnight hair.
Sound of crickets sifting sand.
Marching ants across the land.

Bare feet touch cooling ground,
Cascading water all around.
Heaviness, lifts up high,
Clouds pass, azure sky.

CLIFF PALACE, MESA VERDE N.P., CO.

PENASCO BLANCO, CHACO CULTURE N.H.P., N.M.

BALLCOURT, WUPATKI N.M., AZ.

PREHISTORIC KIVA LADDER, SOUTHEAST UTAH.

 FAJADA BUTTE FROM HUNGO PAVI, CHACO CULTURE N.H.P., N.M.

T-SHAPED DOORWAY, SPRUCE TREE HOUSE, MESA VERDE N.P., CO.

PUEBLO BONITO, CHACO CULTURE N.H.P., N.M.

SYCAMORE & MONTEZUMA CASTLE, MONTEZUMA CASTLE N.M., AZ.

GRANARY, ZION N.P., UT.

PUEBLO BONITO, CHACO CULTURE N.H.P., N.M.

LONG HOUSE, BANDELIER N.M., NEW MEXICO.

PONCHO HOUSE, NAVAJO RESERVATION, UT.

BACK WALL AT CHETTRO KETL, CHACO CULTURE N.H.P., N.M.

LONG HOUSE, MESA VERDE N.P., CO.

STEPS, TSANKAWI SITE, BANDELIER N.M., NEW MEXICO.

DUSK

Color teases the human eye,
Glowing clouds paint the sky.
Ravens caw their last good night,
Winged shadows glide from sight.

Evening cliffs glow amber-red,
Smells waft from baking bread.
Bats swoop at bugs mid-flight.
Coyotes howl. Moon's first light.

Mesquite fires warm evening shade,
Pots a'steam with meals we've made.
Midday warmth soon fades away,
Heated rooms are where we'll stay.

Attend the council, Elder ones,
Teach wisdom to the young.
Pass down our philosophy,
And stimulate curiosity.

Insight that the spirits give,
How it is that we must live.
So tis taught with spoken word.
Is not all life of one accord?

Tell the legends of ancient ones,
Pass stories from old to young.
Exciting hunts of days gone by,
A supernova crossed our sky.

Traveling tales of a distant clan.
Kachinas dancing across the land.
Homes of people with different myths,
Some on hilltops, some in cliffs.

Rhythmic songs by firelight,
Gently soothe a child good night.
Grateful for the day gone by,
Grateful for the night that's nigh.

CLIFF PALACE OVERVIEW, MESA VERDE N.P., CO.

SQUARE TOWER HOUSE, MESA VERDE N.P., CO.

KIVA IN CEREMONIAL CAVE, BANDELIER N.M., NEW MEXICO.

MISSION OF SAN BUENAVENTURA, GRAN QUIVIRA UNIT,
SALINAS PUEBLO MISSIONS N.M., NEW MEXICO.

UPPER & LOWER WHITE HOUSE, CANYON de CHELLY N.M., AZ.

GRANARY, NATURAL BRIDGES N.M., UT.

WUKOKI, WUPATKI N.M., AZ.

RUIN IN SLICKHORN CANYON, GRAND GULCH, UT.

MISSION de NUESTRA SENORA de LOS ANGELES de PORCIUNCULA,
PECOS N.H.P., N.M.

PUEBLO del ARROYO, CHACO CULTURE N.H.P., N.M.

CASA RINCONADA, CHACO CULTURE N.H.P., N.M.

STRONGHOLD HOUSE, HOVENWEEP N.M., UT.

HILLTOP RUIN, GRAND CANYON N.P., AZ.

BETATAKIN, NAVAJO N.M., AZ.

GRANARY, MONUMENT VALLEY TRIBAL PARK, AZ.

LOMAKI, WUPATKI N.M., AZ.

MID-NIGHT

THE CASTLE, HOVENWEEP N.M., UT.

Midnight creatures come to play,
Shadow light is where they stay.
Moonbeams grace this desert land,
Silhouettes a cactus grand.

Scamper out from walls of clay,
Filch the food not put away,
What mice leave, ants take away.
So all is clean by break of day.

Spiritual cause and effect.
(In all things is, deep respect).
It is good, this scheme of things,
To always keep, our Earth clean.

Stars ablaze from where i lie,
Arm to arm across the sky.
i ponder midnight of this race,
Why they went and left this place.

Was it disease, drought or blight,
Made them disappear from sight?
Perhaps the overuse of space,
Caused their move from this place.

How can i learn from them?
Apply it now, what happened then?
Am i wise enough to know,
Lessons learned from long ago?

i listen to the water's song,
Crickets singing all night long.
Circle now these stones of old,
Protect me from the midnight's cold.

The instrument of wind in trees.
At last, i fall asleep with ease.
A haunting question as i sleep,
Shall he that sows, also reap?

PUEBLO del ARROYO, CHACO CULTURE N.H.P., N.M.

JACAL STYLE CONSTRUCTION, CEDAR MESA, UT.

WALNUT CANYON N.M., AZ.

ROADSIDE RUIN, CANYONLANDS N.P., UT.

CASTLE RUIN, GRAND GULCH, UT.

LOWRY RUIN, NORTHERN SAN JUAN BASIN, CO.

THE CITADEL, WUPATKI N.M., AZ.

ATSINNA, EL MORRO N.M., NEW MEXICO.

KIVA & MISSION de NUESTRA SENORA de LOS ANGELES
de PORCIUNCULA, PECOS N.H.P., N.M.

MUMMY CAVE RUIN, CANYON de CHELLY N.M., AZ.

HANDPRINT PICTOGRAPHS & GRANARY, CEDAR MESA, UT.

SQUARE TOWER GROUP, HOVENWEEP N.M., UT.

ANTELOPE HOUSE, CANYON de CHELLY N.M., AZ.

T-SHAPED DOORWAY, CHACO CULTURE N.H.P., N.M.

HORSESHOE RUIN, HOVENWEEP N.M., UT.

THREE ROOM RUIN, SOUTHEAST UTAH.

DAWN ANEW

PUEBLO BONITO, CHACO CULTURE N.H.P., N.M.

Do fallen homes of long ago,
Give the answer we need know.
These broken pieces, this treasure vast,
Will it help us sort the past?

i gaze upon this labored land,
Stacks of stones from Ancient Man.
It's not so hard to hear their song,
And feel their pulse all day long.

To hear the hum of grinding corn,
To know the pain of babies born.
The fight for life from day to day.
It doesn't seem so far away.

i vision tradesmen on these roads,
Distant goods in their loads.
Colored birds from the south,
Painted pots and tales by mouth.

Turquoise from an eastern hill,
A necklace made with ocean shell.
Customs shared by them and those,
From religion to their clothes.

Now tis i here all alone,
Seeing lives etched in stone.
i make past visions, visions mine,
Separate only by length of time.

So i ponder what's to be,
In five hundred years, they'll look at me.
Find my pictures on wooden sheets.
And petroglyphs from past elite.

Will my lifetime win respect?
Or will it only bring regret?
Oneness with Earth, i won't deny.
For its future begins with i.

SELECTED PARKS of the PLATEAU

The National Park Service is charged with the administration, interpretation and preservation of many significant prehistoric cultural sites throughout the Four Corners region. Further information as well as site maps and brochures may be obtained by writing the superintendent.

Arches National Park-125 West 200 South, Moab, UT 84532
Aztec Ruins National Monument-P.O. Box 640, Aztec, NM 87410
Bandelier National Monument-HCR 1, Box 1, Suite 15, Los Alamos, NM 87544
Canyon de Chelly National Monument-P.O. Box 588, Chinle, AZ 86503
Canyonlands National Park-125 West 200 South, Moab, UT 84532
Capitol Reef National Park-Torrey, UT 84775
Chaco Culture National Historical Park-Star Route 4, Box 6500, Bloomfield, NM 87413
El Morro National Monument-Rt. 2, Box 43, Ramah, NM 87321
Glen Canyon National Recreation Area-P.O. Box 1507, Page, AZ 86040
Grand Canyon National Park-P.O. Box 129, Grand Canyon, AZ 86023
Hovenweep National Monument-McElmo Rt., Cortez, CO 81321
Mesa Verde National Park-Mesa Verde National Park, CO 81330
Montezuma Castle National Monument-P.O. Box 219, Camp Verde, AZ 86322
Natural Bridges National Monument-Box 1, Lake Powell, UT. 84533
Navajo National Monument-HC-71, Box 3, Tonalea, AZ 86044
Pecos National Historical Park-P.O. Drawer 11, Pecos, NM 87552
Petrified Forest National Park-P.O. Box 217, Petrified Forest National Park, AZ 86028
Salinas Pueblo Missions National Monument-P.O. Box 496, Mountainair, NM 87036
Tuzigoot National Monument-P.O. Box 219, Camp Verde, AZ 86322
Walnut Canyon National Monument-Walnut Canyon Road, Flagstaff, AZ 86004
Wupatki National Monument-2717 N. Steves Blvd. Suite 3, Flagstaff, AZ 86004
Zion National Park-Springdale, UT 84767-1099

Natural History and Museum Associations, non-profit organizations chartered by Congress to aid the National Park Service, are excellent sources of published material which interpret the Park or Monument in which they operate. Their publications may be purchased at the sales areas in visitor centers or by contacting them directly.

Canyonlands Natural History Association-30 South 100 East, Moab, UT 84532
Capitol Reef Natural History Association-Torrey UT 84775
Glen Canyon Natural History Association-P.O. Box 581, Page, AZ 86040
Grand Canyon Natural History Association-P.O. Box 399, Grand Canyon, AZ 86023
Mesa Verde Museum Association-P.O. Box 38, Mesa Verde National Park, CO 81330
Petrified Forest Museum Association-P.O. Box 277, Petrified Forest National Park, AZ 86028
Southwest Parks and Monuments Association-221 N. Court Avenue, Tucson, AZ 85701
Zion Natural History Association-Zion National Park, Springdale, UT 84767

Our national parks and monuments receive most of the attention but visitors to the region should not overlook the many sites administered by the Bureau of Land Management, U.S. Forest Service and State Parks in Arizona, Colorado, New Mexico and Utah. Information may be obtained by writing.

BUREAU of LAND MANAGEMENT

Anasazi Heritage Center-27501 Highway 184, Dolores, CO 81323
Farmington Resource Area-1235 La Plata Highway, Farmington, NM 87401
Grand Resource Area-885 South Sand Flats Road, Moab, UT 84532
Rio Puerco Resource Area-435 Montano Rd., Albuquerque, NM 87107
San Juan Resource Area-701 Camino del Rio, Durango, CO 81301
Vermillion Resource Area-225 N. Bluff, St. George, UT 84770

U.S. FOREST SERVICE

Coconino National Forest-2323 Greenlaw Ln., Flagstaff, AZ 86001
Dixie National Forest-P.O. Box 580, Cedar City, UT 84720
Manti-La Sal National Forest-599 West Price River Drive, Price, UT 84501
San Juan National Forest-701 Camino del Rio, Durango, CO 81301
Yavapai Visitor Center-P.O. Box 219, Camp Verde, AZ. 86322

STATE PARKS

Arizona State Parks-800 West Washington #415, Phoenix, AZ 85007

New Mexico-Museum of New Mexico Information Office, P.O. Box 2087, Santa Fe, NM 87504
Utah-Division of Parks and Recreation, 1636 West North Temple, Salt Lake City, UT 84116

PRIVATE & TRIBAL ORGANIZATIONS
Crow Canyon Archaeological Center-23390 County Road K, Cortez, CO 81321
Grand Circle Association-P.O. Box 987, Page, AZ 86040
Monument Valley Tribal Park-P.O. Box 308, Window Rock, AZ 86515
San Juan County Archaeological Center-P.O. Box 125, Bloomfield, NM 87413
Santa Clara Pueblo-P.O. Box 580, Española, NM 87532

ABOUT YOUR VISIT

During your visit to the Prehistoric Cultural sites of the Four Corners region, seize the many opportunities to discover and enjoy them but keep in mind that only a small percentage of the surviving sites have been studied or stabilized. Most all, along with thousands of unexcavated sites, possess great potential for further study. As new techniques are developed, more will be learned about these ancient ones, their lifestyle, and causes for the rise and decline of their culture.

Further archaeological study of the prehistoric cultures will only be possible with your assistance. You must demonstrate care, concern and respect by accepting responsibility for your actions and the actions of your group or family. As you visit archeological sites, respect the evidence that has survived for centuries. Walls that remain standing are extremely fragile and pot sherds are fascinating but are only meaningful to researchers if they remain in place.

Petroglyphs (rock etchings) and pictographs (rock paintings) have been the focus of wonder and amazement to countless visitors for decades. Little is known of their meanings or for whom they were intended but we do know that they are extremely fragile. In addition, many of the rock art sites are considered sacred by contemporary Pueblo and Hopi peoples, descendants of the original artists.

Enjoy your visit to these sites but please keep in mind that these sites are an integral part of our national heritage and just as significant as the great chapels, castles, shrines and walls of Europe and Asia. Just as the Vatican and St. Pauls Basilica are sacred to Christianity, these sites are sacred to the descendants of their original inhabitants. Please treat them with the same respect as you would your own sacred relics.

LICHEN ENCRUSTED PETROGLYPH, LITTLE BLACK MTN., AZ.

PHOTOGRAPHIC CREDITS

Mary Allen: 10.
Russ Bishop: 59.
Gunnar Conrad: 87.
Michael Fatali: 25.
Eduardo Fuss: 23, 24, 60.
Fred Hirschmann: 9, 70, 83.
Bruce Hucko: 11, 16, 17, 18, 19, 20, 29, 42, 52, 53, 69, 80, 82, 91.
George H. H. Huey: 5, 14, 30, 31, 46, 51, 58, 71, 85.
Lewis Kemper: 86.
Mark Miller: 8.
Jeff Nicholas: 6, 12, 13, 15, 22, 27, 33, 40, 43, 48, 49, 55, 62, 64, 66 68, 81, 84, 88, 90, 95.
Greg Probst: 41.
Norm Shrewsbury: 72.
Tom Till: 26, 36, 38, 44, 73, 77, 78.
Larry Ulrich: 37, 65, 76.
Jim Wilson: Cover, 1, 7, 28, 32, 34, 35, 45, 47, 50, 54, 56, 61, 63, 67, 74, 79, 89, 92.

We would like to take this opportunity to extend our sincere appreciation to all those photographers who submitted work for consideration on this project. The overall quality of work, and our limited needs, meant a heartbreaking number of extraordinary images were left out of the final edit (much to our frustation!).

CREDITS

Introduction by Jeff Nicholas
"The Prehistoric Drama" by Florence C. Lister
"Windows" by Lynn Wilson
"Parks of the Colorado Plateau" compiled by Jim Wilson
"About Your Visit" by Jim Wilson
Comparative Time-Line by Jeff Nicholas

Editor-Nicky Leach
Design-Jeff Nicholas
Typesetting by MacinType, Fresno, Ca.
Separations, printing and binding by Tien Wah Press, Ltd., Singapore

SUGGESTED READING

Banham, P. Reyner. *Scenes in America Deserta.* (1982). Reprint. Cambridge: MIT Press, 1989.

Leach, Nicky. *The Guide to the National Parks of the Southwest.* Tucson: Southwest Parks and Monuments Assoc., 1992.

Lister, Florence C. and Robert H. Lister. *Those Who Came Before.* (1983). Reprint. Tucson: Southwest Parks and Monuments Assoc. 1989.

Lopez, Barry H. *Crossing Open Ground.* New York: Macmillan Publishing Co. 1988.

Lopez, Barry H. *The Rediscovery of North America.* Lexington: University of Kentucky Press. 1990.

Noble, David Grant. *Ancient Ruins of the Southwest.* (1981). Reprint. Flagstaff: Northland Publishing Co. 1991.

Ortiz, Alfonso (Editor). *The Handbook of North American Indians, Volume 9.* Washington, DC: Smithsonian Institution. 1979.

Schaafsma, Polly. *Indian Rock Art of the Southwest.* (1980). Reprint. Albuquerque: University of New Mexico Press. 1990.

Schaafsma, Polly. *Rock Art in New Mexico.* Santa Fe: Museum of New Mexico Press. 1992.

Snyder, Gary. *Turtle Island.* (1969). Reprint. New York: New Directions Books. 1974.

Waters, Frank. *The Book of the Hopi.* (1963). Reprint. New York: Penguin Books. 1982.

A wide range of books, pamphlets and trail guides are usually available at very affordable prices from the Natural History or Museum Association affiliated with the Park they serve. The list of associations on page 94 is included to make acquisition of detailed information easier for the traveler planning to visit these sites.